SCHOLASTIC

GRADES K-2

PLAY & LEARN MATH
Place Value
Learning Games and Activities to Help Build Foundational Math Skills

by Mary Rosenberg

New York • Toronto • London • Auckland • Sydney
Mexico City • New Delhi • Hong Kong • Buenos Aires

Scholastic Inc. grants teachers permission to print and photocopy the reproducible pages from this book for classroom use. Purchase of this book entitles use of reproducibles by one teacher for one classroom only. No other part of this publication may be reproduced in whole or in part, or stored in a retrieval system, or transmitted in any form or by any means, electronic, mechanical, photocopying, recording, or otherwise, without written permission of the publisher. For information regarding permission, write to Scholastic Inc., 557 Broadway, New York, NY 10012.

Editor: Maria L. Chang
Cover design by Tannaz Fassihi
Cover art by Constanza Basaluzzo
Interior design by Grafica Inc.
Interior art by Mike Moran

Scholastic Inc., 557 Broadway, New York, NY 10012
ISBN: 978-1-338-28562-8
Copyright © 2019 by Mary Rosenberg
All rights reserved.
Printed in the U.S.A.
First printing, January 2019.
1 2 3 4 5 6 7 8 9 10   131   25 24 23 22 21 20 19

Contents

Introduction 5
Mathematics Standards Correlations 6

Circle a Number 7
Blocks in a Cup 9
By the Scoop 11
Snaps and Claps 14
Build a Number 16
Place Value Trees 19
How Many Ways? 26
By the Tens 28
Skip Counting 31
Place Value Riddles 37
Which Is Greater? 40
Show Me the Number 43
Compare the Numbers 46
Tens and Hundreds 51
Record the Number 55
More Skip Counting 57
More or Less 59
It's in the Cards 62

Introduction

Place value is the foundation for learning in mathematics. Studying place value teaches children what numbers really mean. They learn that each digit in a number represents a certain value depending upon its place (location). A 5 in the hundreds place is different from a 5 in the tens place and from a 5 in the ones place. In the hundreds place, 5 equals 500, but in the tens place, 5 means 50. In the ones place, 5 is simply 5. Place value helps children understand that 65 is greater than 56 and that $30 is worth more than $3.

Children develop a solid understanding of place value through activities that focus on **decomposing numbers** (breaking numbers apart into hundreds, tens, and ones) and **composing numbers** (putting numbers together). By playing with numbers—decomposing and composing numbers, showing the same amount in different ways, and using manipulatives—children get a sense of how different arithmetic operations work. This enables them to have an easier time learning addition, subtraction, multiplication, and division.

Play & Learn Math: Place Value offers lots of fun opportunities to play with numbers. The activities are aligned with the Common Core State Standards in Mathematics with a focus on Numbers and Operations in Base Ten (NBT) for kindergarten through 2nd grade. (See page 6 for a matrix showing the standards supported by each activity.) However, you can use the activities with any grade level based upon the specific needs of each child.

Many of the activities in this book are written in two ways: one to guide teachers on how to implement the activity, and another to give children instructions using kid-friendly language. The purpose of this is simple—the teacher introduces and models the activity to the whole class or to small groups, then children can use their own copy, written at their level, as a step-by-step reminder of how to do the activity. The activities require only a few simple classroom materials, such as place value blocks, number cubes (dice), and playing cards, and are ideal for math centers and small groups. Children can work independently or with a partner to improve their skills and develop concept understanding.

Let's get playing!

Mathematics Standards Correlations*

Standard	Circle a Number	Blocks in a Cup	By the Scoop	Snaps and Claps	Build a Number	Place Value Trees	How Many Ways?	By the Tens	Skip Counting	Place Value Riddles	Which Is Greater?	Show Me the Number	Compare the Numbers	Tens and Hundreds	Record the Number	More Skip Counting	More or Less	It's in the Cards
KINDERGARTEN																		
CC.A.1 Count to 100 by ones and by tens.		✓	✓						✓							✓		
CC.A.3 Write numbers from 0-20. Represent a number of objects with written numeral 0-20.	✓	✓	✓						✓									
CC.B.4 Understand the relationship between numbers and quantities.		✓	✓										✓					
CC.C.6 Identify whether the number of objects in one group is greater than, less than, or equal to the number of objects in another group.			✓									✓	✓					
CC.C.7 Compare two numbers between 1 and 10 presented as written numerals.												✓	✓					✓
NBT.A.1 Compose and decompose numbers from 11 to 19 into ten ones and some further ones (e.g., by using objects or drawings), and record each composition or decomposition by a drawing or equations (e.g., 18 = 10 + 8); understand that these numbers are composed of ten ones and one, two, three, four, five, six, seven, eight, or nine ones.	✓	✓	✓	✓	✓	✓	✓											
GRADE 1																		
NBT.A.1 Count to 120, starting at any number less than 120. Read and write numerals and represent a number of objects with a written numeral.									✓			✓						
NBT.B.2 Understand that the two digits of a two-digit number represent amounts of tens and ones.	✓	✓	✓	✓	✓	✓	✓	✓	✓	✓	✓	✓	✓	✓				
NBT.B.2.A 10 can be thought of as a bundle of ten ones.	✓	✓	✓	✓	✓	✓	✓	✓	✓	✓								
NBT.B.2.B The numbers from 11 to 19 are composed of a ten and one, two, three, four, five, six, seven, eight, or nine ones.	✓	✓	✓	✓	✓	✓	✓					✓						
NBT.B.2.C The numbers 10, 20, 30, 40, 50, 60, 70, 80, 90 refer to one, two, three, four, five, six, seven, eight, or nine tens (and 0 ones).				✓	✓	✓	✓	✓	✓			✓	✓	✓				
NBT.B.3 Compare two two-digit numbers based on meanings of the tens and ones digits, recording the results of comparisons with the symbols >, =, and <.											✓		✓					✓
GRADE 2																		
NBT.A.1 Understand that the three digits of a three-digit number represent amounts of hundreds, tens, and ones.						✓	✓	✓	✓			✓	✓	✓	✓			
NBT.A.1.A 100 can be thought of as a bundle of 10 tens.						✓	✓	✓						✓				
NBT.A.1.B The numbers 100, 200, 300, 400, 500, 600, 700, 800, 900 refer to one, two, three, four, five, six, seven, eight, or nine hundreds (and 0 tens and 0 ones).						✓	✓	✓				✓	✓	✓				
NBT.A.2 Count within 1,000; skip-count by 5s, 10s, and 100s.									✓							✓		
NBT.A.3 Read and write numbers to 1,000 using base-ten numerals, number names, and expanded form.						✓				✓		✓			✓			
NBT.A.4 Compare two three-digit numbers using >, =, and < symbols to record the results of comparisons.											✓		✓	✓				✓
NBT.B.5 Fluently add and subtract within 100 using strategies based on place value, properties of operations, and/or the relationship between addition and subtraction.						✓												✓
NBT.B.6 Add up to four two-digit numbers using strategies based on place value and properties of operations.																		✓
NBT.B.7 Add and subtract within 1,000, using concrete models or drawings and strategies based on place value, properties of operations, and/or the relationship between addition and subtraction.																		✓
NBT.B.8 Mentally add 10 or 100 to a given number 100-900, and mentally subtract 10 or 100 from a given number 100-900.																	✓	

* © 2010 National Governors Association Center for Best Practices and Council of Chief State School Officers. All rights reserved.

Teacher Page

Circle a Number

Assess how well children know numbers and place value with this introductory activity. Children pick a number from 11 to 19 and represent the number in different ways—by writing the number word or the numeral, by drawing, by completing a number sentence, and so on.

HERE'S HOW

Distribute copies of the "Circle a Number" activity page to children. Display a copy on the board.

Guide children on how to complete the activity page. On their sheet, have children pick and circle one of the numbers from 11 to 19. Then have them write either the numeral or the number word to complete the sentence. (Provide children with Number Words cards, as needed.) Next, have children color in the tens blocks to show their number. Then have them complete the number sentence and write the number in the place value chart.

For example:

MATERIALS

- Circle a Number activity page (page 8)
- crayons
- pencils
- Number Words cards for 11–19 (page 45), optional
- classroom projection system

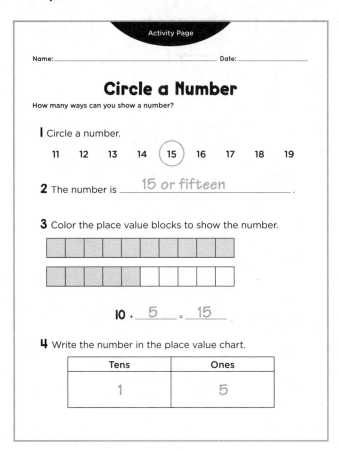

Circle a Number

How many ways can you show a number?

1 Circle a number.

11 12 13 14 15 16 17 18 19

2 The number is _____ .

3 Color the place value blocks to show the number.

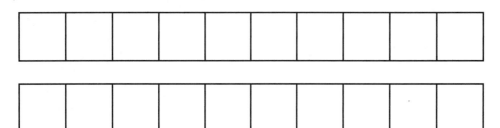

10 + _____ = _____

4 Write the number in the place value chart.

Tens	Ones

Teacher Page

Blocks in a Cup

This activity relates counting to numbers and place value. After counting the number of blocks in a cup, children figure out how many tens and ones make up the number. Children can work independently or in pairs. A perfect addition to your math learning center.

HERE'S HOW

For each table, prepare 10 cups labeled A to J. Place 11 to 19 ones place value blocks in each cup in random order.

Distribute copies of the "Blocks in a Cup" activity page to children. Display a copy on the board.

Model how to do the activity. Pick a cup and record the cup's letter on the activity page. Shake the cup and pour the blocks onto the table. Count the blocks. On the page, write how many blocks were in the cup and color in the corresponding number of blocks in the place value chart. Next, sort the blocks into groups of tens and ones. Finally, write the number on the page.

Have children do the activity on their own or with a partner.

MATERIALS

- 10 large paper or plastic cups, labeled A, B, C, D, E, F, G, H, I, J
- place value blocks (ones)
- Blocks in a Cup activity page (page 10)
- pencils
- crayons
- classroom projection system

Activity Page

Name: _____ Date: _____

Blocks in a Cup

How many blocks are in a cup?

What to Do

1 Pick a cup from your teacher. Write the letter on the cup below.

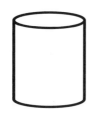

You'll Need
- ★ labeled cups from your teacher
- ★ pencil

2 Shake the cup. Pour the blocks on the table. Count the blocks. Color in the number of blocks below.

☐☐☐☐☐☐☐☐☐☐
☐☐☐☐☐☐☐☐☐☐

There were _____ blocks in the cup.

3 Sort the blocks into groups of tens and ones.

Tens	Ones
☐☐☐☐☐☐☐☐☐☐	☐☐☐☐☐☐☐☐☐

4 There were _____ tens and _____ ones in the cup.

5 The number is _____.

Teacher Page

By the Scoop

This is another great math learning center activity. Children use different sizes of measuring cups to scoop ones blocks from a basket. They then count the number of blocks and decompose the number into tens and ones. This also presents the perfect opportunity for children to compare quantities in different measuring cups and understand which holds more. Children can work individually or in pairs.

HERE'S HOW

Distribute copies of "By the Scoop" recording sheet to children. Display a copy on the board.

Model how to do the activity. Select a measuring cup and scoop up the place value blocks. Make as many tens as possible with the blocks. On the recording sheet, draw the tens in the Tens space. (Show children how to make a skinny rectangle to represent the tens. Younger children might want to draw the lines to show the individual ones in each ten.) Draw the ones in the Ones space. Finally, record the number made in the Number section. Children can write the numeral or the number word or use expanded form (e.g., 43, forty-three, 40 + 3, 4 tens 3 ones).

Have children do the activity on their own or with a partner, using different-size measuring cups.

MATERIALS

- By the Scoop recording sheet (page 13)
- dry measuring cups in $\frac{1}{4}$, $\frac{1}{3}$, $\frac{1}{2}$, and 1-cup sizes
- large basket of ones place value blocks
- pencils
- classroom projection system

Independent Practice

By the Scoop

Scoop up some blocks with a measuring cup. How many blocks did you get?

What to Do

1. Pick a measuring cup. Use it to scoop up the place value blocks.

2. Make as many tens as possible with the blocks.

3. On your recording sheet, draw the tens in the Tens space. Draw the ones in the Ones space.

4. Write the number you made in the Number space.

5. Do Steps 1 to 4 again. Use a different-size measuring cup.

You'll Need

★ By the Scoop recording sheet

★ different sizes of measuring cups

★ large basket of ones place value blocks

★ pencil

Name: _____ Date: _____

By the Scoop

Scoop Size	Tens	Ones	Number
1/4 cup			
1/3 cup			
1/2 cup			
1 cup			

Teacher Page

Snaps and Claps

This activity can be done as a whole class or in small groups. For each snap of the fingers you make, children place a ones block on the Ones side of their place value mat. Once they have 10 blocks, they clap their hands to exchange the 10 ones for a tens block. A great introduction to the concept of regrouping.

HERE'S HOW

Distribute copies of the "Snaps and Claps" place value mat and tens and ones place value blocks to children. (Each child should have at least 10 ones blocks.)

Explain to children that their place value mat is divided into two sections: Tens and Ones. Each time you snap your fingers, they should place one place value block (single ones) on the Ones side of the mat. They should then name the number they made. For example:

0 tens 1 one = 1

Tens	Ones
	□

0 tens 2 ones = 2

Tens	Ones
	□ □

When children have 10 blocks on the Ones side of their place value mat, they should clap their hands. They can then exchange their 10 ones for a tens block (or a rod), and place it on the Tens side of their mat. Children should continue naming the number they make.

For example:

1 ten 0 ones = 10

Tens	Ones
□□□□□□□□□□	

Continue snapping and clapping until you reach a desired number.

MATERIALS

- Snaps and Claps place value mat (page 15)
- place value blocks (ones and tens)
- number cube, optional

VARIATION

Once children have had a chance to do the activity several times, you can replace snapping and clapping fingers with rolling a number cube for a faster-paced game. Children can play this game with a partner or in a small group with three or four players. Start by picking a target number. Players take turns rolling the number cube and placing the matching number of ones blocks on his or her place value mat. As with the activity, once they have 10 blocks on the Ones side, children should exchange them for a tens block and place it on the Tens side. Players continue taking turns until someone reaches the target number and wins.

Place Value Mat

Name: _____ Date: _____

Snaps and Claps

Listen as your teacher snaps his or her fingers. For each snap, place one block under Ones. When you have 10 blocks, clap your hands. Then swap your 10 ones blocks for one tens block. Place the tens block under Tens side.

Tens	Ones

Teacher Page

Build a Number

Children show their understanding of place value by representing a two- or three-digit number in two different ways. This activity helps them gain a solid understanding that 10 ones is the same as 1 ten and that 10 tens is the same as 1 hundred—an important concept in addition, subtraction, and other operations.

HERE'S HOW

For kindergarten and 1st grade, distribute copies of the "Build a Number: Tens and Ones" activity page and provide children with tens and ones blocks.

For 2nd grade, distribute copies of the "Build a Number: Hundreds, Tens, and Ones" activity page and provide children with hundreds, tens, and ones blocks.

Guide children on how to do the activity. Select a two-digit number (or a three-digit number for older children). Using place value blocks, have them show two different ways to make that number on their activity page. Then have them complete the sentence frame describing how many of each place value block they used to make the number.

For example:

MATERIALS

- Build a Number activity pages (pages 17–18)
- place value blocks (ones, tens, and hundreds)
- pencils

Activity Page #1

Name: _____ Date: _____

Build a Number: Tens and Ones

I can show the number __19__ two different ways.

Way #1 Using Only Ones	Way #2 Using Tens and Ones
(19 ones blocks)	(1 ten block and 9 ones blocks)
I can show the number __19__ using __19__ ones.	I can show the number __19__ using __1__ tens and __9__ ones.

Name: _____ Date: _____

Activity Page #1

Build a Number: Tens and Ones

I can show the number _____ two different ways.

Way #1
Using Only Ones

Way #2
Using Tens and Ones

I can show the number _____ using _____ ones.

I can show the number _____ using _____ tens and _____ ones.

Name: _____ Date: _____

Activity Page #2

Build a Number: Hundreds, Tens, and Ones

I can show the number _____ two different ways.

Way #1
Using Only Tens and Ones

Way #2
Using Hundreds, Tens, and Ones

I can show the number _____ using _____ tens and _____ ones.

I can show the number _____ using _____ hundreds, _____ tens, and _____ ones.

Teacher Page

Place Value Trees

In this activity, children practice decomposing and composing numbers. Picking a random number, children decompose it by drawing how many tens and ones make up the number. They then compose the number by adding the place values, which is the same as writing a number in expanded form.

HERE'S HOW

Photocopy and cut apart a set of level-appropriate Number Cards for each child or pair of children. Place the cards in a paper bag. Also, photocopy the appropriate "Place Value Trees" recording sheet for each child. Display a copy on the board.

Model how to do the activity. Pick a number card from the paper bag and write the number in the top box of the tree. Draw the number of tens and ones (or hundreds, tens, and ones) represented by the number (see Note, right). Then write the number in expanded form in the spaces below.

Have children work on their own or in pairs to complete their recording sheet.

For example:

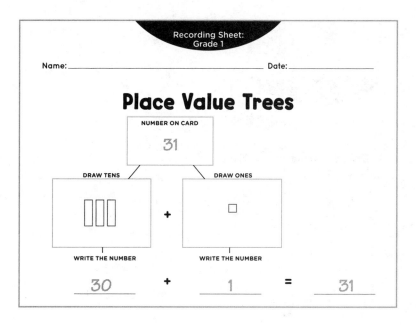

MATERIALS

- **Number Cards** (pages 21–22)
- **Place Value Trees recording sheets** (pages 23–25)
- paper bag
- classroom projection system

NOTE: Show children how to draw a quick picture of the place value blocks.

☐ = hundreds

▯ = tens

▫ = ones

Independent Practice

Place Value Trees

Pick a number. Then fill in the Place Value Tree.

What to Do

1 Pick a number card. Write the number in the top box.

2 Draw the number of tens and ones (or hundreds, tens, and ones).

☐ = hundreds | = tens   ☐ = ones

3 Write the number in expanded form.

4 Do Steps 1 to 3 again.

You'll Need
★ Number Cards
★ Place Value Trees recording sheet
★ pencil

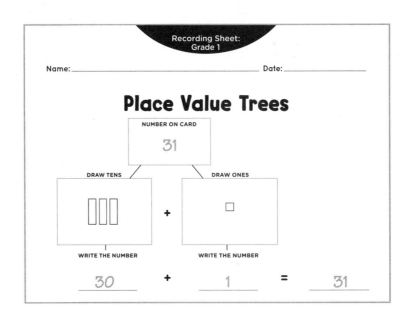

Number Cards

For Kindergarten

11	12	13
14	15	16
17	18	19

For 1st Grade

25	31	47
56	63	82
99	105	120

Number Cards

For 2nd Grade

128	160	172
257	319	345
491	500	634
783	817	946

Recording Sheet:
Grade K

Name: _____ Date: _____

Place Value Trees

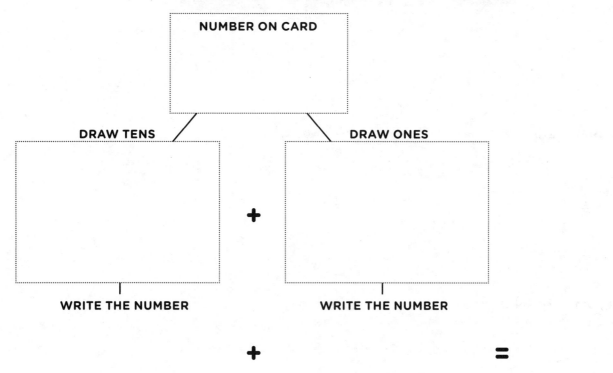

NUMBER ON CARD

DRAW TENS   DRAW ONES

+

WRITE THE NUMBER   WRITE THE NUMBER

_____ + _____ = _____

NUMBER ON CARD

DRAW TENS   DRAW ONES

+

WRITE THE NUMBER   WRITE THE NUMBER

_____ + _____ = _____

Recording Sheet: Grade 1

Name: _____ Date: _____

Place Value Trees

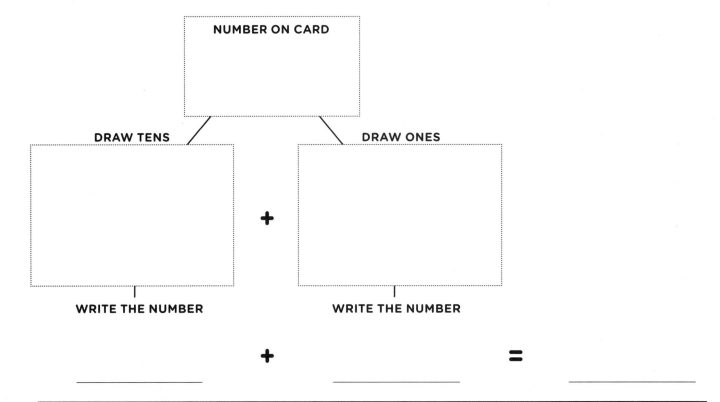

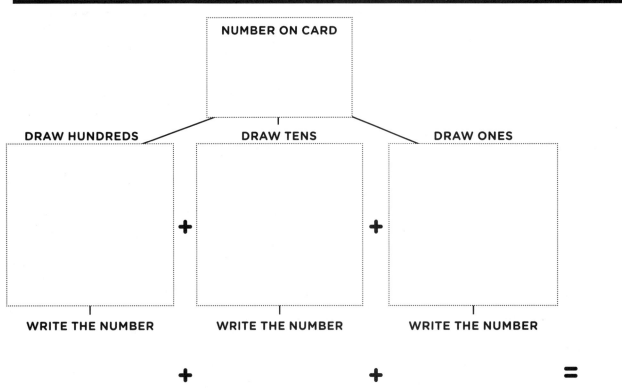

Recording Sheet: Grade 2

Name: _____ Date: _____

Place Value Trees

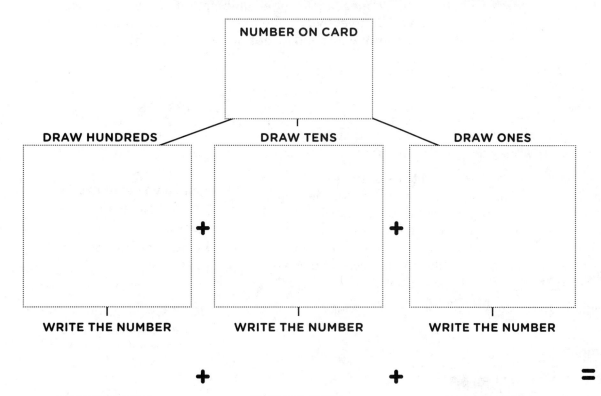

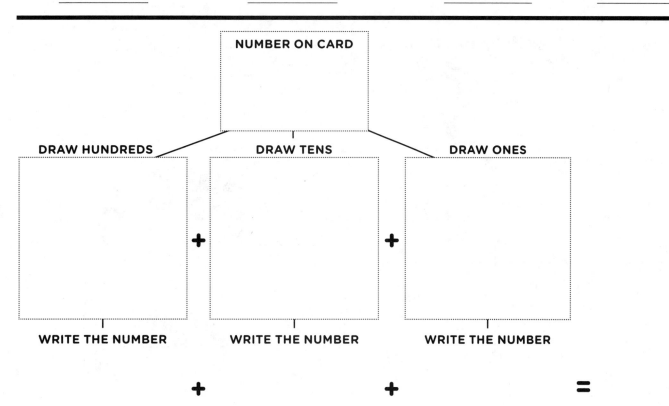

Teacher Page

How Many Ways?

This activity offers children flexibility in how they show a given number using place value blocks. They can show a number using all ones blocks, or a combination of tens and ones, or of hundreds, tens, and ones blocks. While the activity page has space for four ways to show a number, keep in mind that with some numbers there may not be four ways to show it.

HERE'S HOW

Distribute copies of the "How Many Ways?" activity page to children. Display a copy on the board.

Model how to do the activity. Pick a two- or three-digit "target number" and write it in the center of the dartboard. (You can call out a number for the whole class or for each child, or have children choose their own number.) Using place value blocks, show the number in as many different ways as possible. Then, draw the place value blocks on the chart (see Note, right).

Have children do the activity on their own or with a partner.

For example:

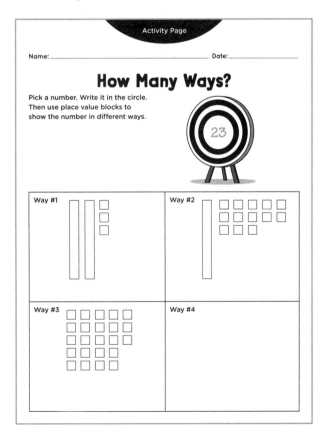

MATERIALS

- How Many Ways? activity page (page 27)
- place value blocks (ones, tens, and hundreds)
- pencils
- classroom projection system

NOTE: Show children how to draw a quick picture of the place value blocks.

Activity Page

Name: _____ Date: _____

How Many Ways?

Pick a number. Write it in the circle. Then use place value blocks to show the number in different ways.

Way #1	Way #2
Way #3	Way #4

Teacher Page

By the Tens

Children work in pairs in this activity, which focuses on the tens place value. They use place value blocks to represent each number, then compare and discuss with their partner how they made their numbers.

HERE'S HOW

Partner up children. Give each child his or her own copy of the "By the Tens" activity page.

Tell children to use place value blocks to show each number on their activity page. Afterwards, have them compare how they made their numbers with their partner. Did partners make each number the same way, or did they use different blocks?

Challenge children to show the same number using different place value blocks. For example, 20 can be shown using 2 tens, or 1 ten and 10 ones, or 20 ones.

MATERIALS

- By the Tens activity page (page 30)
- place value blocks (ones and tens)

Independent Practice

By the Tens

Use place value blocks to show the tens. Then compare with your partner.

What to Do

1. Show each number on your activity page. Use place value blocks.

2. Compare your work with your partner. Did you make each number the same way? Or did you use different blocks?

3. Do Step 1 again. Use different combinations of place value blocks.

You'll Need
- ★ By the Tens activity page
- ★ place value blocks
- ★ a partner

Name: _____ Date: _____

Activity Page

By the Tens

50 fifty	**100** one hundred	**70** seventy	**90** ninety	**40** forty
80 eighty	**30** thirty	**10** ten	**20** twenty	**60** sixty

Teacher Page

Skip Counting

Using place value blocks, children investigate which way of counting is more efficient—by ones or by twos, and later by fives or by tens. They then extend this learning by coloring in or filling in hundred charts while skip counting.

HERE'S HOW

Give each child two copies of the "Skip Counting" activity page, a pencil, and place value blocks. Display a copy of the activity page on the board.

Ask children to predict: *Which do you think is a faster way of counting—by ones or by twos?* On one activity page, have children put a ones place value block in each circle and write the ongoing total number below each circle. Model on the board by drawing the blocks and writing the numbers.

Next, have children take their second activity page and place two blocks in each circle. Have them write the ongoing total number below each circle. When everyone has finished, invite some children to count by ones to 20 while other children count by twos. Ask: *Which way is faster?* (Counting by twos)

Afterwards, distribute the complete "120 Chart" (page 33) to children. Have them count by twos and color in the appropriate numbers. Alternatively, you can give children the "Skip Counting by 2s" chart (page 34) and have them fill in the blank spaces while counting by twos.

At a later date, repeat the activity, skip counting by 5s and by 10s (use tens place value blocks).

MATERIALS

- Skip Counting activity page (page 32)
- place value blocks (ones and tens)
- pencils
- Hundred Charts (pages 33-36)
- crayons
- classroom projection system

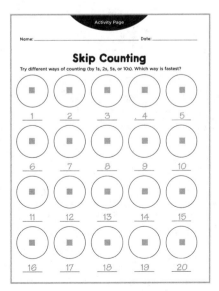
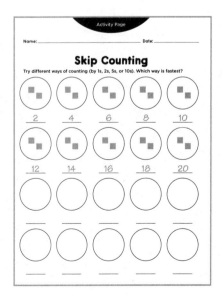

Activity Page

Name: _____ Date: _____

Skip Counting

Try different ways of counting (by 1s, 2s, 5s, or 10s). Which way is fastest?

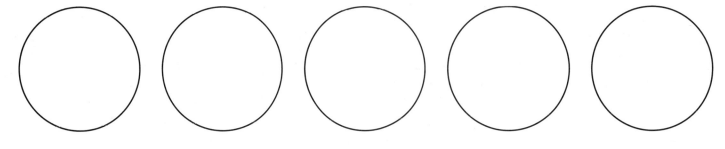
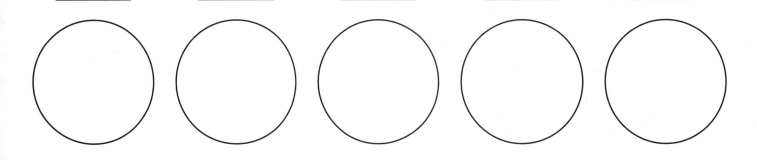
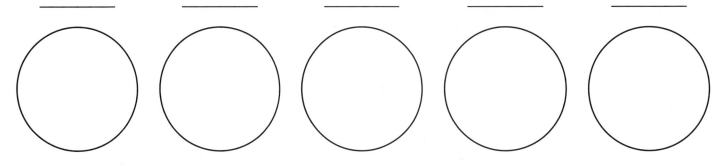
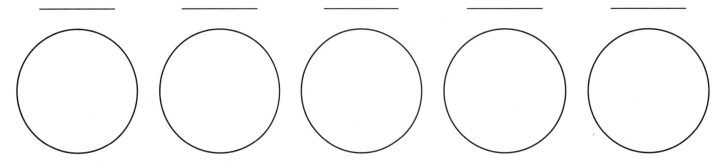

120 Chart

1	2	3	4	5	6	7	8	9	10
11	12	13	14	15	16	17	18	19	20
21	22	23	24	25	26	27	28	29	30
31	32	33	34	35	36	37	38	39	40
41	42	43	44	45	46	47	48	49	50
51	52	53	54	55	56	57	58	59	60
61	62	63	64	65	66	67	68	69	70
71	72	73	74	75	76	77	78	79	80
81	82	83	84	85	86	87	88	89	90
91	92	93	94	95	96	97	98	99	100
101	102	103	104	105	106	107	108	109	110
111	112	113	114	115	116	117	118	119	120

Skip Counting by 2s

Fill in the missing numbers. Say the numbers as you write them.

1		3		5		7		9	
11		13		15		17		19	
21		23		25		27		29	
31		33		35		37		39	
41		43		45		47		49	
51		53		55		57		59	
61		63		65		67		69	
71		73		75		77		79	
81		83		85		87		89	
91		93		95		97		99	
101		103		105		107		109	
111		113		115		117		119	

Hundred Chart #3

Name: _____ Date: _____

Skip Counting by 5s

Fill in the missing numbers. Say the numbers as you write them.

1	2	3	4		6	7	8	9	
11	12	13	14		16	17	18	19	
21	22	23	24		26	27	28	29	
31	32	33	34		36	37	38	39	
41	42	43	44		46	47	48	49	
51	52	53	54		56	57	58	59	
61	62	63	64		66	67	68	69	
71	72	73	74		76	77	78	79	
81	82	83	84		86	87	88	89	
91	92	93	94		96	97	98	99	
101	102	103	104		106	107	108	109	
111	112	113	114		116	117	118	119	

Skip Counting by 10s

Fill in the missing numbers. Say the numbers as you write them.

1	2	3	4	5	6	7	8	9	
11	12	13	14	15	16	17	18	19	
21	22	23	24	25	26	27	28	29	
31	32	33	34	35	36	37	38	39	
41	42	43	44	45	46	47	48	49	
51	52	53	54	55	56	57	58	59	
61	62	63	64	65	66	67	68	69	
71	72	73	74	75	76	77	78	79	
81	82	83	84	85	86	87	88	89	
91	92	93	94	95	96	97	98	99	
101	102	103	104	105	106	107	108	109	
111	112	113	114	115	116	117	118	119	

Teacher Page

Place Value Riddles

Play this riddle game with the whole class or in small groups. Children identify the number and use place value blocks to show the number. They then show the number four other ways: expanded form, standard form, number word, and tens and ones. Afterwards, invite children to create their own place value riddles for classmates to solve.

HERE'S HOW

Distribute copies of the "Place Value Riddles" recording sheet to children.

Pick a riddle to read aloud. Have children show their answer using place value blocks on their recording sheets. Then have them show the same number four other ways. Repeat as many times as you wish.

For example:

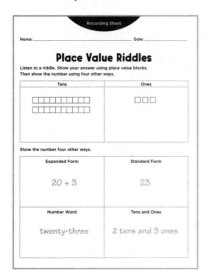

Afterwards, invite children to write their own place value riddles. Give each child a blank sentence strip on which to write his or her riddle. Then have children exchange riddles with one another and draw a picture on the strip to show their answer.

MATERIALS

- Place Value Riddles recording sheet* (page 38)
- pencils or dry-erase or water-based markers
- Place Value Riddles cards (page 39)
- place value blocks (ones and tens)
- blank sentence strips or index cards

* You may want to laminate the recording sheet or put copies in plastic sheet protectors. This way, children can write on them with dry-erase markers (or water-based markers for sheet protectors). They can then easily erase their writing with a rag between riddles.

Answers to riddles are shown below in bold print.

I have 2 tens and 3 ones. What number am I? **23**	I have 8 ones and 4 tens. What number am I? **48**	I have 3 tens and the same number of ones. What number am I? **33**
I have 6 ones and 1 ten. What number am I? **16**	I have 9 ones and 0 tens. What number am I? **9**	I have 2 tens and 0 ones. What number am I? **20**
I have 5 ones and the same number of tens. What number am I? **55**	I have 4 ones and 6 tens. What number am I? **64**	I have 7 tens and 1 one. What number am I? **71**
I have 8 tens and 2 ones. What number am I? **82**	I have 5 ones and 1 ten. What number am I? **15**	I have 2 tens and the same number of ones. What number am I? **22**
I have 3 ones and 6 tens. What number am I? **63**	I have 5 ones and 3 tens. What number am I? **35**	I have 4 tens and 9 ones. What number am I? **49**

Recording Sheet

Name: _____ Date: _____

Place Value Riddles

Listen to a riddle. Show your answer using place value blocks. Then show the number using four other ways.

Tens	Ones

Show the number four other ways.

Expanded Form	Standard Form
Number Word	Tens and Ones

38

Riddle Cards

Place Value Riddles

I have 2 tens and 3 ones. What number am I?	I have 8 ones and 4 tens. What number am I?	I have 3 tens and the same number of ones. What number am I?
I have 6 ones and 1 ten. What number am I?	I have 9 ones and 0 tens. What number am I?	I have 2 tens and 0 ones. What number am I?
I have 5 ones and the same number of tens. What number am I?	I have 4 ones and 6 tens. What number am I?	I have 7 tens and 1 one. What number am I?
I have 8 tens and 2 ones. What number am I?	I have 5 ones and 1 ten. What number am I?	I have 2 tens and the same number of ones. What number am I?
I have 3 ones and 6 tens. What number am I?	I have 5 ones and 3 tens. What number am I?	I have 4 tens and 9 ones. What number am I?

Teacher Page

Which Is Greater?

In this activity, children pick two numbers at random, represent them using place value blocks, then compare them using the appropriate symbol (<, >, or =). Having a physical representation of the numbers helps children visualize quantity and determine which number is greater or less than the other. You can have children do this activity independently or with a partner.

HERE'S HOW

Distribute copies of the "Which Is Greater?" recording sheet to children. Give each child or pair of children the appropriate number cards (two- or three-digit numbers or a combination of both), stacking them facedown on the table.

Display a copy of the recording sheet on the board and model how to do the activity. Turn over a card and write the number on the left side of the recording sheet. Then turn over a second card and write the number on the right side of the sheet. Use the place value blocks to show each number. Then write the appropriate symbol (<, >, or =) in the space between the numbers. Fill in the sentence frames and read aloud the inequality (or equality).

Have children repeat the activity a few times.

MATERIALS

- **Which Is Greater? recording sheet*** (page 42)
- pencils or dry-erase or water-based markers
- **Number Cards** (pages 21–22, 50)
- place value blocks (ones, tens, and hundreds)
- classroom projection system

* You may want to laminate the recording sheet or put copies in plastic sheet protectors. This way, children can write on them with dry-erase markers (or water-based markers for sheet protectors), which they can easily erase with a rag.

Independent Practice

Which Is Greater?

Pick two numbers and compare them.

What to Do

1. Place the number cards facedown on the table.

2. Turn over a card. Write the number on the left side of your recording sheet.

3. Turn over another card. Write the number on the right side of your sheet.

4. Use the place value blocks to show each number.

5. Compare the numbers. Write the symbol (<, >, or =) between the numbers.

6. Fill in the sentences. Then read them aloud.

7. Do Steps 1 to 6 again a few more times.

You'll Need

★ Which Is Greater? recording sheet
★ Number Cards
★ place value blocks
★ pencil or dry-erase marker

Name: _____ Date: _____

Recording Sheet

Which Is Greater?

Card #1		Card #2	
Symbol		Symbol	
Place Value Blocks		Place Value Blocks	

_____ is greater than _____.

_____ is less than _____.

Show Me the Number

While still using place value blocks, children explore other ways to represent numbers, including standard form, number words, expanded form, and more. A great way to reinforce number sense and place value.

HERE'S HOW

Distribute copies of the "Show Me the Number" activity page to children. Give each child three level-appropriate number cards.

Display a copy of the activity page on the board. Go over the example with children. Point out how each section of the chart is completed. Ask children to identify the various sections of the chart. For example: *Which shows a number word? Which one uses place value blocks? How is the expanded form different from the tens and ones form?*

Have children use their number cards to fill in their activity page. Provide Number Words cards, as needed. You can have children work individually or with a partner.

If there's time, invite children to share one of their numbers and the different ways they showed it.

MATERIALS

- Show Me the Number activity page (page 44)
- Number Cards (pages 21–22, 50)
- place value blocks (ones, tens, and hundreds)
- pencils
- Number Words cards (page 45), optional
- classroom projection system

NOTE: Show children how to draw a quick picture of the place value blocks.

☐ = hundreds

❘ = tens

▫ = ones

Name: _____ Date: _____

Activity Page

Show Me the Number

Pick three number cards. Write the numbers under Standard Form. Then fill in the rest of the chart.

Standard Form	Place Value Blocks	Words	Expanded Form	Tens and Ones	Another Way
Example: 87	(8 tens rods, 7 ones)	Eighty-seven	80 + 7	8 tens 7 ones	(7 tens rods, 17 ones)

Number Words

0	zero				
1	one	11	eleven	21	twenty-one
2	two	12	twelve	22	twenty-two
3	three	13	thirteen	30	thirty
4	four	14	fourteen	40	forty
5	five	15	fifteen	50	fifty
6	six	16	sixteen	60	sixty
7	seven	17	seventeen	70	seventy
8	eight	18	eighteen	80	eighty
9	nine	19	nineteen	90	ninety
10	ten	20	twenty	100	one hundred

Teacher Page

Compare the Numbers

This activity comes with three activity pages, each increasing in difficulty. Children compare ones, tens, up to hundreds. You can use these pages to assess children's understanding of place value or to differentiate for your students.

HERE'S HOW

Distribute the appropriate "Compare the Numbers" activity page to children.

For the "Tens and Hundreds" page, model how to compare numbers by looking at the digits in the largest place first.

- If the digits in the hundreds place are different from each other, then compare those particular numbers.

    For example: 385 and 425

    3 (300) is less than 4 (400), so 385 is less than 425.

- If both digits in the hundreds place are the same, then move to the right and compare the digits in the tens place.

    For example: 385 and 325

    Both digits in the hundreds place are the same (300), so compare the digits in the tens place. 8 (80) is greater than 2 (20), so 385 is greater than 325.

- If both digits in the tens place are the same, move to the right again and compare the digits in the ones place.

    For example: 385 and 382

    Both digits in the hundreds place are the same, as are the digits in the tens place. So compare the digits in the ones place. 5 is greater than 2, so 385 is greater than 382.

Have children complete the activity pages on their own or with a partner.

MATERIALS

- **Compare the Numbers activity pages*** (pages 47–49)
- **Number Cards** (page 50)
- pencils

*Note that "Compare the Numbers: Tens and Hundreds" (page 49) comes with Number Cards.

Answers

Compare the Numbers: Ones (page 47)
1. 8 **2.** 9 **3.** 5 < 8; 5 is less than 8 **4.** 4 > 0; 4 is greater than 0 **5 and 6.** Answers will vary.

Compare the Numbers: Tens (page 48)
1. 30 < 50; 30 is less than 50 **2.** 80 > 40; 80 is greater than 40 **3.** 90 < 100; 90 is less than 100 **4.** 70 > 60; 70 is greater than 60 **5–8.** Answers will vary.

Compare the Numbers: Tens and Hundreds (page 49)
Answers will vary.

Activity Page #1

Name: _____ Date: _____

Compare the Numbers: Ones

Count the ones. Compare the numbers. Circle the greater amount.

1. ☐☐☐☐   ☐☐☐
   ☐☐☐☐   ☐☐☐

   _____   _____

2. ☐☐☐☐☐   ☐☐☐☐
   ☐☐☐☐

   _____   _____

Compare the numbers. Fill in the blanks. Use > (greater than) or < (less than).

3. ☐☐☐☐☐   ☐☐☐☐
              ☐☐☐☐

   5 ____ 8

   5 is _____ than 8.

4. ☐☐☐☐

   4 ____ 0

   4 is _____ than 0.

5. Draw more than 5 ones.
   How many ones did you draw? _____ ones

6. Draw less than 8 ones.
   How many ones did you draw? _____ ones

Activity Page #2

Name: _____   Date: _____

Compare the Numbers: Tens

Compare the numbers. Fill in the blanks. Use > (greater than) or < (less than).

1.

   30 _____ 50

   30 is _____ 50.

2.

   80 _____ 40

   80 is _____ 40.

3.

   90 _____ 100

   90 is _____ 100.

4.

   70 _____ 60

   70 is _____ 60.

Draw the tens to make each number sentence true.

5.

   80 > _____

   80 is greater than _____ .

6.

   50 < _____

   50 is less than _____ .

7.

   10 < _____

   10 is less than _____ .

8.

   60 > _____

   60 is greater than _____ .

Compare the Numbers: Tens and Hundreds

Pick two number cards. Draw the place value blocks. Compare the numbers.

Example:

____33____ < ____41____

____33____ is ____less____ than ____41____.

1.

____ ◯ ____

____ is ____ than ____.

2.

____ ◯ ____

____ is ____ than ____.

3.

____ ◯ ____

____ is ____ than ____.

4.

____ ◯ ____

____ is ____ than ____.

5.

____ ◯ ____

____ is ____ than ____.

Number Cards

Tens

41	28	17
50	33	56
48	15	25
79	87	62
72	81	94

Hundreds

117	434	374
556	258	921
795	529	942
885	670	836
701	418	167

Teacher Page

Tens and Hundreds

Children gain an understanding that just as 10 ones make 1 ten, 10 tens make 1 hundred. The three activity pages increase in difficulty, so you can use them to scaffold learning or to differentiate for your students.

HERE'S HOW

Distribute the appropriate "Tens and Hundreds" activity page to children. Display a copy on the board.

Using the example on each sheet, model for children how to complete each activity page. For the #2 and #3 activity pages, show children how to draw a quick picture of the place value blocks (see below).

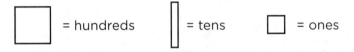

Have children complete the activity pages on their own or with a partner. Monitor them as they work and provide help, as needed.

Answers

Tens and Hundreds #1 (page 52)
1. 8 tens; 0 hundreds 8 tens; 80 **2.** 15 tens; 1 hundred 5 tens; 150
3. 20 tens; 2 hundreds 0 tens; 200 **4.** 10 tens; 1 hundred 0 tens; 100
5. 19 tens; 1 hundred 9 tens; 190

Tens and Hundreds #2 (page 53)
1. 13 tens **2.** 20 tens **3.** 30 tens **4.** 21 tens **5.** 24 tens

Tens and Hundreds #3 (page 54)
1. 1 hundred and 3 tens **2.** 12 tens **3.** 25 tens **4.** 20 tens
5. 2 hundreds 4 tens

MATERIALS

- Tens and Hundreds activity pages (pages 52–54)
- pencils
- classroom projection system

Name: _____   Date: _____

Tens and Hundreds: Circle the Tens

Count the tens. Write how many tens. Circle 10 tens to make a hundred. Record the number of hundreds and tens. Write the total number.

Example:	_12_ tens
	_1_ hundreds _2_ tens
	_120_
1.	____ tens
	____ hundreds ____ tens
	____
2.	____ tens
	____ hundreds ____ tens
	____
3.	____ tens
	____ hundreds ____ tens
	____
4.	____ tens
	____ hundreds ____ tens
	____
5.	____ tens
	____ hundreds ____ tens
	____

Activity Page #1

Activity Page #2

Name: _____ Date: _____

Tens and Hundreds: Draw the Tens

Show each number. Draw the tens. How many tens did you make?

Example: 150

150 = 15 tens

1. 130

130 = _____ tens

2. 200

200 = _____ tens

3. 300

300 = _____ tens

4. 210

210 = _____ tens

5. 240

240 = _____ tens

Activity Page #3

Name: _____ Date: _____

Tens and Hundreds: Which Is More?

Draw the tens to show each amount. Circle the larger number.

Example:	18 tens	(1 hundred 9 tens)
1.	1 hundred 3 tens	10 tens
2.	12 tens	1 hundred
3.	2 hundreds	25 tens
4.	20 tens	1 hundred 8 tens
5.	2 hundreds 4 tens	22 tens

Teacher Page

Record the Number

In this activity, a chart lists several three-digit numbers in one of five different ways: standard form; number name; place value blocks; hundreds, tens, and ones; and expanded form. Children have to rewrite those numbers using the other ways. A great way to reinforce place value up to hundreds.

HERE'S HOW

Distribute copies of the "Record the Number" activity page to children. Display a copy on the board.

Model how to fill in the chart. Point out that each row gives one way to show a number. The task is to show that same number in different ways, according to the heading in each column.

Have children complete the chart on their own or with a partner. Allow them to use place value blocks and the Number Words cards, as needed.

NOTE: Show children how to draw a quick picture of the place value blocks.

☐ = hundreds | = tens   ▫ = ones

MATERIALS

- Record the Number activity page (page 56)
- pencils
- place value blocks (ones, tens, hundreds)
- Number Words cards (page 45), optional
- classroom projection system

Answers

Standard Form	Number Name	Place Value Blocks	Hundreds, Tens, and Ones	Expanded Form
394	**three hundred ninety-four**		3 hundreds 9 tens 4 ones	300 + 90 + 4
757	seven hundred fifty-seven		**7 hundreds 5 tens 7 ones**	700 + 50 + 7
271	two hundred seventy-one		2 hundreds 7 tens 1 one	**200 + 70 + 1**
892	eight hundred ninety-two		8 hundreds 9 tens 2 ones	800 + 90 + 2
450	four hundred fifty		4 hundreds 5 tens 0 ones	400 + 50 + 0

Name: _____ Date: _____

Activity Sheet

Record the Number

Complete the chart. Show each number in different ways.

Standard Form	Number Name	Place Value Blocks	Hundreds, Tens, and Ones	Expanded Form
	three hundred ninety-four			
			7 hundreds 5 tens 7 ones	
				200 + 70 + 1
892				

Teacher Page

More Skip Counting

Skip counting is an essential skill that's useful in telling time, counting money, and solving addition, subtraction, multiplication, and division problems. Have children practice skip counting throughout the day—while lining up, walking in the hallways, at the end of the day, and so on. Children should practice counting not only forwards but also backwards. It is also important that children master skip counting with numbers that do not end in 0 or 5; for example, have them skip count by 5s starting at 11 or 112 or 329.

HERE'S HOW

Distribute copies of the "More Skip Counting" activity page to children.

Explain to children that they'll be skip counting by 5s, 10s, and 100s—both forwards and backwards. Have them fill in the missing numbers for each counting pattern. Tell them to pay close attention to the patterns to see if the numbers are going up or down.

Answers

1. 20, 25, 30 **2.** 140, 150, 165 **3.** 815, 820, 825 **4.** 60, 55, 40
5. 70, 80, 90 **6.** 240, 250, 280 **7.** 720, 690, 670 **8.** 980, 960, 940
9. 330, 430, 530 **10.** 590, 790, 990 **11.** 900, 800, 600
12. 560, 460, 260

MATERIALS

- More Skip Counting activity page (page 58)
- pencils

More Skip Counting

Fill in the blanks to complete each pattern.

Skip Counting by 5s

1. 5, 10, 15, _____, _____, _____

2. 135, _____, 145, _____, 155, 160, _____

3. 800, 805, 810, _____, _____, _____

4. 75, 70, 65, _____, _____, 50, 45, _____

Skip Counting by 10s

5. 50, 60, _____, _____, _____, 100

6. 210, 220, 230, _____, _____, 260, 270, _____, 290

7. 730, _____, 710, 700, _____, 680, _____, 660

8. 1000, 990, _____, 970, _____, 950, _____, 930

Skip Counting by 100s

9. 130, 230, _____, _____, _____, 630

10. 390, 490, _____, 690, _____, 890, _____

11. 1000, _____, _____, 700, _____, 500

12. 660, _____, _____, 360, _____, 160

Teacher Page

More or Less

Children should be able to mentally add 10 and 100 to any number as well as subtract 10 and 100 from that same number. This activity provides practice for this important math skill. Using number cubes gives randomness to the activity, helping develop automaticity.

HERE'S HOW

Distribute copies of the "More or Less" recording sheet and number cubes to children. Display a copy on the board.

Model how to do the activity. Roll three number cubes to make a three-digit number. Choose which digit goes in each place value. (Alternatively, you can roll one number cube three times.) Record the number on the first column of recording sheet. Then fill in the rest of that row by adding 10 more and 100 more to the number and taking away 10 and 100 from the number.

Have children do the activity on their own or with a partner. Each child should fill in his or her own chart.

MATERIALS

- More or Less recording sheet (page 61)
- number cubes
- pencils
- classroom projection system

Independent Practice

More or Less

**Add 10 and 100 to a number.
Then subtract 10 and 100 from the same number.**

What to Do

1 Make a three-digit number. Roll the number cubes. (Decide which digit goes in each place value.)

2 Record the number under Number Rolled.

3 Fill in the rest of the chart. Add 10 more and 100 more to the number. Then take away 10 and 100 from the number.

4 Do Steps 1 to 3 four more times.

You'll Need
★ More or Less recording sheet
★ 3 number cubes
★ pencil

Name: _____ Date: _____

Recording Sheet

More or Less

Number Rolled	10 More	100 More	10 Less	100 Less
EXAMPLE: 682	692	782	672	582

Teacher Page

It's in the Cards

In this two-player card game, children compare numbers at each turn. They also get practice adding multi-digit numbers. After the game, encourage children to discuss what they've noticed while playing. For example, if a player has the greater one-digit and two-digit numbers, does she automatically get the greater sum?

HERE'S HOW

Distribute copies of the "It's in the Cards" recording sheet to children. Display a copy on the board. Model how to play this two-player card game:

1. Deal 10 cards to each player. The players lay the cards facedown in front of them to form a pyramid (see right).

2. Both players turn over their top card and record the number on their recording sheets—both on the pyramid and in the Adding Section. The player with the greater number wins the round and draws a star next to the number.

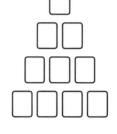

3. Both players turn over the cards in the second row and record the two-digit number on the pyramid and in the Adding Section. The player with the greater two-digit number wins the round and draws a star next to the number.

4. The players then add the numbers in the first and second row and record the sum. The player with the greater sum wins the round and places a star next to the number.

5. Play continues as described in Steps 2 to 4, until all the cards have been turned over. The player with the most stars wins.

MATERIALS

- It's in the Cards recording sheet (page 64)
- set of playing cards with the 10s, jacks, queens, and kings removed (for each pair of children)
- pencils
- classroom projection system

Game Directions

It's in the Cards

Compare and add numbers with this two-player card game.

What to Do

1. Shuffle the cards. Deal 10 cards to each player.

2. Lay your cards facedown in front of you. Form a pyramid, as shown below.

3. Turn over your top card. On your recording sheet, write the number on the pyramid and in the Adding Section. The player with the greater number draws a star next to the number. That player wins the round.

4. Turn over the cards in the second row. On your recording sheet, write the two-digit number in the pyramid and in the Adding Section. The player with the greater number draws a star next to the number. That player wins the round.

5. Add the numbers in the first and second row of the Adding Section. Record the sum. The player with the greater sum draws a star next to the number. That player wins the round.

6. Play continues with the third and fourth row of cards. Write the numbers, compare them, and add them.

7. The player with the most stars wins.

Players: 2

You'll Need

★ It's in the Cards recording sheet
★ set of playing cards
★ pencil

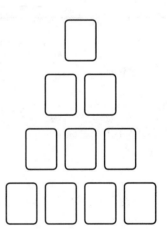

Recording Sheet

Name: _____ Date: _____

It's in the Cards

First Row

Second Row

Third Row

Fourth Row

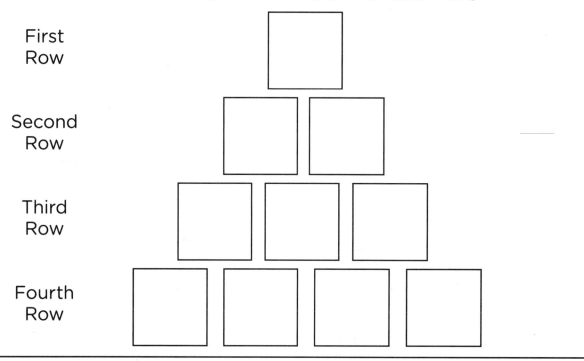

Adding Section

First Row

Second Row

+

Sum

Third Row

+

Sum

Fourth Row

+

Sum

64